Bridging Scientific Knowledge:
The Role of the Medical Science Liaison

Susan E. Malecha

authorHOUSE®

AuthorHouse™
1663 Liberty Drive, Suite 200
Bloomington, IN 47403
www.authorhouse.com
Phone: 1-800-839-8640

First published by AuthorHouse 2/23/2009

ISBN: 978-1-4343-9681-5 (sc)

Library of Congress Control Number: 2009900385

Printed in the United States of America
Bloomington, Indiana

This book is printed on acid-free paper.

For Jim

Table of Contents

Introduction

The pharmaceutical industry offers a variety of jobs in many technical areas. As in any industry, there is a multitude of titles, departments, and responsibilities, many of which are unfamiliar or confusing to those outside of the industry. One of these positions was recently touted by CNN.com as one of the best jobs a person could have: the medical science liaison (http://money.cnn.com/galleries/2007/news/0702/gallery.jobs_in_demand/index.html).

For those who work as a medical science liaison (MSL), it can be a challenge to clearly define the position. When a MSL is asked what he/she does, the initial reply might be: "I'm a field-based medical science liaison for Company X."

The inquirer will most likely reply, "So, you're a sales rep?"

To which the MSL replies, "No, I am not a sales rep. I'm a medical science liaison." The subsequent thirty-second explanation then turns into a mini-lesson about a role in the pharmaceutical/biotechnology industry.

Although there are similarities between sales representatives and MSLs in that both are members of field-based geographic teams, the two roles serve very different functions. The purpose of the sales force is to produce sales, and this involves the art of personal selling. This salesmanship involves a standardized process of prospecting and qualifying the customer, the approach, presentation and demonstration, overcoming objections, closing, and follow-up and maintenance. Also, for a sales rep, an important aspect of the position is negotiation, the art of arriving at a sales transaction with terms that satisfy both parties. In contrast, an MSL does not sell a product, and the notion of a sales objective and related compensation is irrelevant. The responsibility that both sales reps and MSLs do have in common, however, is relationship management: the art of creating a closer working relationship and interdependence between the people in two organizations

Most people assume that biotech or pharmaceutical employees who are not based at corporate headquarters are sales representatives. This is not the case. One of the first concepts to grasp about the work of an MSL is the unique job location, which is field-based. In other words, the individual works from a home office, not at corporate headquarters, and has responsibilities for a designated geographical area. The travel requirements of an MSL depend on the size of the assigned territory, the goals/responsibilities of the position, and the needs of the business.

Another characteristic of an MSL's job is that it involves a large amount of face-to-face interaction. This is not an office desk job, or one for the extreme introvert. On the contrary, this position involves frequent one-on-one communication with health care professionals, researchers, and many other health care professionals.

MSLs are flourishing in today's pharmaceutical and biotechnology industries, and they play a variety of roles within their organizations. Most MSL positions involve dissemination of timely, complex scientific information; development of educational programs; communication with opinion leaders; and training and facilitation of clinical research/investigation.

Because each pharma/biotech company has its own unique business objective for the MSL, the programs and activities of an MSL can take various strategic directions. However, their clinical credibility and ability to communicate peer-to-peer is something that all MSLs have in common. This affords their companies the ability to meet the increasingly sophisticated informational needs of customers.

Most MSLs have a background as a health care professional and have an advanced degree. For example, many MSLs currently in the industry have earned a Doctor of Pharmacy (Pharm.D.), Doctor of Philosophy (Ph.D.), or a Medical Doctor (M.D.) degree. In many instances, the role is filled by nurses, most of whom have an advanced degree. While many pharmaceutical companies require advanced degrees, it is often the case that a qualified MSL does not have such a degree, but possesses extensive industry or clinical experience in a given therapeutic area.

The purpose of this book is to provide a general overview of the MSL's role in the pharmaceutical/biotech industry. For those working in this industry, it is hoped that this book will clarify the role of MSLs and why they are so important to the scientific communications of a biotech/pharma company. This book also seeks to increase understanding of the MSL role among professionals who collaborate with the industry including physicians, clinical researchers, managed care representatives, and others in the health care professional arena.

This book will cover three distinct areas:
- Medical Science Liaison Role and Responsibilities. Defines the purpose and business value that the MSL position brings to scientific communication and research development.
- Medical Science Liaison Operations. Reviews the activities of the MSL, the handling of proprietary and confidential information within regulatory guidelines, training needs, and activities that support the business.
- Medical Science Liaison and Investigator-Sponsored Trials. Defines a specific type of clinical study, describes the process by which such studies are supported, and explains MSLs' impact on this process.

Note:
Depending on the company, the medical science liaison may alternatively be called a medical sciences manager, clinical trial liaison, clinical manager, field-based liaison, or research liaison. However, for consistency, this book will refer to the position as a medical science liaison (or MSL).

Chapter 1:
Medical Science Liaison Role and Responsibilities

One of the responsibilities of the medical science liaison is to establish and facilitate research collaborations between the company and a research investigator or health care professional. An MSL must therefore be able to meet a current or prospective researcher face to face, address questions about the company's research goals and product pipeline, and be capable of facilitating the processes that the researcher needs to follow in order to initiate and maintain a collaboration, whether short-lived or ongoing. What makes an MSL more than mere "technical support" or even a "super-sales rep" is their ability to engage in meaningful scientific discussions with the researcher or health care professional, about the disease state as well as the scientific evidence regarding the use of a particular therapeutic intervention. Medical science liaisons engage in peer-to-peer discussions with investigators, and are able to challenge and cultivate scientific ideas in an open dialogue. While a medical science liaison does not solicit research ideas, he/she serves as a channel through which research proposals are submitted to the company. Active scientific exchange leads to the unveiling of questions and ultimately to research projects aimed at elucidating pathways contributing to physiological phenomena—specifically, in the company's therapeutic areas. Such research projects can lead to discoveries that ultimately impact patients' lives.

Examples of clinical support activities of the MSL may include:

- Assisting the clinical department by nominating sites for registration studies
- Assisting the clinical department by attending trial awareness programs and answering questions from investigators
- Assisting in recruitment for registration studies
- Assisting in clinical site referrals
- Assisting investigators in the development of investigator-sponsored clinical research, per corporate strategy

MSL programs have an integral place in the life-cycle development of a product within a therapeutic area. MSLs support the company's commercial direction and are critical to the positive positioning of its therapeutic capability in a given geography, at a given stage in the product life cycle. The provision of complex scientific information, upon request, is a critical function of the MSL. The medical science liaison imparts education and knowledge in various venues, not only in the form of individual responses to requests, but also in data presentations at advisory boards and in their responses to inquiries at the company's scientific conference booths. The trust that is established in an MSL-customer relationship results from consistent demonstration of scientific expertise and satisfactory follow-through to requests.

MSLs have other responsibilities within the organization, and they interact with other functioning teams. In some instances, they may participate in scientific training for company personnel. In some organizations, MSLs help prepare speakers for advisory boards or train contracted speakers about the therapeutic data. Some MSLs work extensively with the clinical development departments, and may facilitate investigator meetings as part of a company clinical trial.

An utmost topmost priority of a MSL is cultivating relationships with regional thought leaders, also called opinion leaders. An opinion leader is:

- An individual with a regional or national "sphere of influence"
- A person with expertise who is well respected in their field

- A person who is guiding the research and influencing the direction that medicine will take in a specific therapeutic area

Since opinion leaders are widely disseminated geographically, MSLs are distributed across the nation to serve as points of contact between scientific researchers and industry.

The MSL can enhance the medical community's awareness of scientific information. In particular, the collaboration between MSLs and opinion leaders can lead toward a greater understanding of a disease state, and a more complete dissection of the roles of classes of drug compounds in the treatment of a disease state. This collaboration can ultimately contribute to the scientific body of evidence for a biological pathway, and bring about a heightened understanding of the current standard of care and how to align medical programs with current practices.

Medical science liaisons actively engage in many activities that support the strategic direction of a product. Consequently, MSLs are exposed to key decision makers, both within the organization and outside of the company. As channels of collaboration between pharmaceutical companies and thought leaders, MSLs are essential to the quality and successful transmission of timely information, research resources, and business intelligence.

Regulations and compliance guidelines govern the activities of the sales force, as well as MSLs. For example, the FDA has guidance for medical communications which apply to MSL activities. Each pharma/biotech company operates within their interpretation of these guidelines. The sales force can only proactively speak about the attributes of the drug that are approved by the FDA.

Chapter 2:
Medical Science Liaison Operations

A Day in the Life of a Medical Science Liaison

Every day is different for a medical science liaison. Some days are consumed with long-distance travel, while other days involve one or more appointments with local investigators. Items discussed during investigator visits may range from clinical trials to disease education activities, depending on the investigator's interests and expressed need for information. For example, one day may include a scheduled visit with a physician to discuss an investigator-initiated research proposal. Such a conversation might typically last anywhere from thirty minutes to an hour or more, covering topics such as primary endpoints, inclusion and exclusion criteria, and efficacy parameters. The study budget may also be addressed. Early on in this conversation, the MSL should describe the company process for reviewing such proposals, including the use of a committee, and should then explain what steps are required to submit a proposal. It is good practice for both individuals to leave the discussion with follow-up action items, such as protocol development or research. The next scheduled appointment of the day might take place with a study coordinator, to check on the status or progress of a company-sponsored or investigator-initiated study. On another day, the MSL may be asked to give a short educational presentation to health care staff, perhaps focusing on recent data from a company-sponsored study or disease education. Alternatively, the medical science liaison may be asked to facilitate a clinical trial awareness activity, such as a lunch or dinner program.

Another typical activity in the life of a MSL may involve attendance at a national conference, such as the American Heart Association Annual Scientific Sessions. At such conferences, the MSL has the opportunity to learn cutting-edge information about one or more disease states while meeting some of the top thought leaders in the therapeutic area. Some MSLs spend time in the conference exhibit hall, staffing the medical section of their company's informational booth. This affords them the opportunity to discuss with health care professionals their company's product, trials, or relevant disease state. Such discussions should only occur within the context of answering unsolicited questions; they must not be construed as promoting the company's products. Days spent at conferences also provide opportunities for the MSL to collect intelligence on competitior products. Conference days frequently end with dinners that the MSL has scheduled with physicians or other health care professionals.

Although it is commonplace for medical science liaisons to conduct their work individually, some days are consumed with mandatory company conference calls, and others with team meetings held at the corporate office. It is important for MSLs to stay abreast of scientific advances in their therapeutic fields. Thus, some time must be focused on learning, be it through online training, in-person seminars, company presentations, or fellowship activities (usually some combination of the above). Lastly, at least some days in the MSL's calendar are administrative, devoted to the preparation of expense reports, activity reports, and other miscellaneous work.

Regardless of the activity of the day, MSLs must always remember that they operate in a highly regulated industry. In other words, MSLs must follow rules governing their interactions with health care professionals, particularly those rules that relate to exchanging medical information and answering unsolicited questions. It is also important for the MSL to remember that there may be differences in the way many companies, and individuals, interpret these guidelines.

Regulations and Guidelines

It is the responsibility of all medical science liaisons to become familiar with the federal regulations and guidances for medical communications. Federal regulations are contained in the *Code of Federal Regulations* (CFR), which is the codification of the general and permanent rules published in the *Federal Register* by the executive departments and agencies of the federal government. Medical communications are addressed in chapter 21 of the CFR. Other sources of guidance regarding the pharmaceutical industry and the conduct of health care professionals, including medical science liaisons, are as follows:

• *PhRMA*: The Pharmaceutical Research and Manufacturers of America (PhRMA) is an organization that "represents the leading research-based pharmaceutical and biotechnology companies in the United States" (http://www.phrma.org/about_phrma/). This organization established a code for interacting with the medical community. One of its purposes is to provide guidelines regarding the pharmaceutical industry's relationships with physicians and other health care professionals. The PhRMA Code aims to support the intention that interactions with health care professionals are to benefit patients and to enhance the practice of medicine. The code also provides that no grants, scholarships, subsidies, support, consulting contracts, or educational or practice-related items should be provided or offered to a health care professional in exchange for prescribing products or for a commitment to continue prescribing products. A full explanation of the code can be found at: http://www.phrma.org/files/PhRMA%20Code.pdf

• *The Accreditation Council for Continuing Medical Education (ACCME) Standards for Commercial Support:*

The ACCME is an organization that identifies, develops, and promotes standards for quality continuing medical education (CME) utilized by physicians in their maintenance of competence and incorporation of new knowledge to improve

quality medical care for patients and their communities. In their policies for commercial support, appropriate courses of action are offered relating to interactions of health care professionals with CME providers and other health care professionals.
http://www.accme.org/index.cfm/fa/Policy.policy/Policy_id/9456ae6f-61b5-4e80-a330-7d85d5e68421.cfm

- *Health and Human Services Office of the Inspector General (OIG):* The OIG provides pharmaceutical and biotechnology manufacturers with compliance program guidance and addresses many topics relevant to health care professional communication. The OIG's resources include fraud alerts, advisory opinions, and information about safe harbor regulations.
  http://oig.hhs.gov/authorities/docs/03/050503FRCPGpharmac.pdf

- Two guidance documents from the American Medical Association address accepted practices for giving gifts to physicians, as well as other related issues. Although medical science liaisons, in general, do not give gifts to physicians or health care providers, there may be an instance when a medical book or other office-related educational item is provided. It is important to understand the rules and guidance governing such exchanges.
  http://www.ama-assn.org/ama/pub/category/5689.html with the clarification addendum
  http://www.ama-assn.org/ama/upload/mm/369/gifts_clarification.pdf

Although medical science liaisons carry out a wide variety of activities and interactions with health care professionals, the regulations and guidelines described above apply to all of their work.

What Tools are Needed to be a Successful Medical Science Liaison? First and Foremost: Training

Training plans for MSLs are complicated but necessary. Because MSLs disseminate cutting-edge medical and technical information, continued training is paramount to their success and credibility. MSL training requires time, energy, planning, and resources.

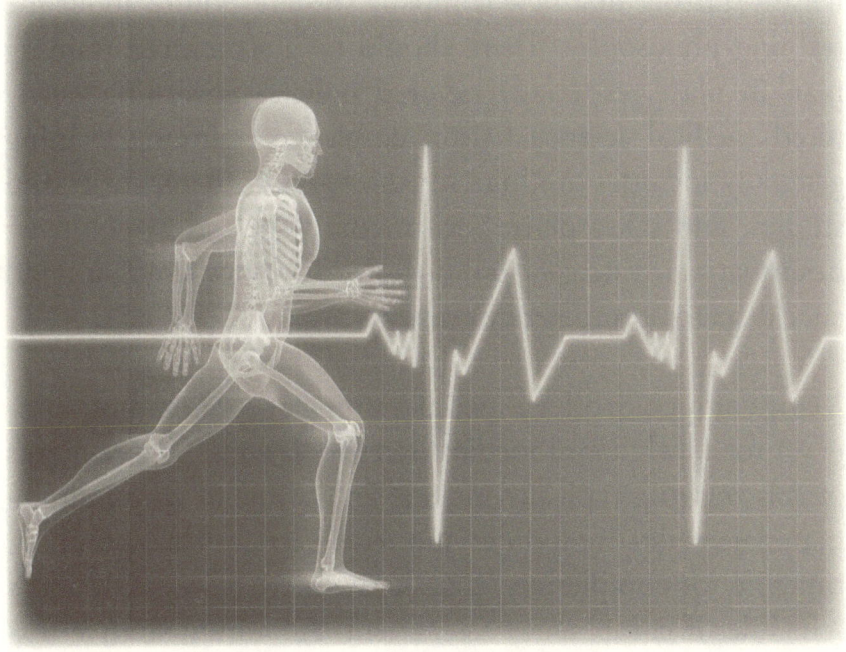

There are risks associated with neglecting MSL training. Many federally mandated regulations, as mentioned previously, affect MSLs. Specifically, the FDA addresses the sci-

entific exchange of information, pharmacovigilance requirements, and Health Insurance Portability and Accountability
Act (HIPAA) regulations. Significant consequences can result
from an MSL inadvertently sharing the wrong information
at the wrong time. Not knowing and understanding the
regulations can have serious consequences.

The two keys to success for an MSL—expertise in a therapeutic area and the ability to communicate—can be fostered
by extensive training. A thorough understanding of patients
and medical management dynamics, managed care and insurance issues, biotech or pharma industry operations and
guidelines, and the ability to effectively intertwine and communicate these issues are paramount. Every medical science
liaison starting in a new company should inquire about the
training philosophy. There should be a structured training
plan in place, especially tailored policies governing newly
hired medical science liaison employee training. Training
generally includes modules for the relevant therapeutics and
studies, and takes place both in the field and at company
headquarters. Effective training also includes timelines and
checklists for competency evaluations.

Many companies offer a special training or "onboarding" program for newly hired medical science liaisons. The elements
of this training typically include an outline of core training
materials, series of modules, summaries of studies, a proposed
timeline for completion (usually six to twelve weeks), and a
schedule of competency assessment. All administrative (expense reports, calendar management, monthly reports, etc.)
and computer training is typically covered in an onboarding
program.

Further components of training usually include a number of important sessions. These can include, at a minimum:

- Services offered by human resources
- Code of conduct and commercial policies
- Function of medical information department
- Drug safety of company products; drug safety department
- Regulatory guidelines
- Function of corporate accounts
- Marketing overview
- Reviews of slide decks focusing on the company's products and relevant disease state
- Frequently asked questions
- Medical science liaison lectures/clinical presentations

Understanding the MSL role within the organization is the first step in mapping competencies against this job function; defining and leveling competencies are one of the several roadmaps that MSLs use to provide consistent service to meet the business objective. These involve both internal customers and external customers. Common MSL competencies include:

- Strategic thinking
- Therapeutic knowledge
- Technology skills
- Negotiation skills
- Regulatory knowledge
- Innovation/creativity
- Business knowledge

- Project management
- Building and leading teams
- Collaboration and teamwork
- Communication
- Decision making
- Results driven

Often, the company clearly defines each competency that is expected of its MSLs, and delineates levels of skill within each competency; e.g., beginner/fundamental, intermediate/moderate, advanced, expert/thought leader.

Training and Development of an MSL

The assumption of any new role requires learning many steps to achieve productive results. Expectations need to be set and communicated, roles clarified, and a focus on skills, knowledge, and experience must be taken into account. Coaching and orientation programs are also important.

Training and development of MSL employees correlates directly with performance management. The employee needs assessment determines the current level of competency. There are five phases to an individual's training and development plan: learning the role, seeking feedback and coaching, developing at a level, developing coaching and mentoring skills, and growing and stretching.

Many techniques help new MSLs to understand their role, such as peer-to-peer interactions, mentoring, and catering the training to the individual need. The MSL should progress through all levels, as the MSL's understanding of the position and their value grow.

Although much of the training is done at home, the MSL will travel to company headquarters for parts of the training. It is important for the MSL to meet important internal employees (clinical development, medical directors, medical communications, for example) and establish relationships with key business stakeholders.

The new MSL will also receive training in the field at some determined point of their onboarding training with their manager. Initially, field training may be described as an "introductory" phase to the business and territory. The goal of this introductory phase is for the medical science liaison to become familiar with key customers (investigators, thought leaders, for example) and the roles, meet key customers in the area, and learn about the territory.

Sample Initial Training Schedule	
Week	*Training to be Completed*
Weeks 1–4	Core Therapeutic Training
Weeks 5–8	Additional Therapeutic Training, Field Training
Weeks 9–11	In-House Training
Week 12	Field Travel

While initial training usually continues for twelve weeks, and sometimes even longer, MSLs are expected to participate in advanced training on an ongoing basis. To continue their own professional development, the MSL needs to be aware of possible areas of growth, and should work with her/his manager to make sure areas are covered. In addition, many MSL teams have a dedicated trainer, who plays an integral role in updating and revising future training materials and resources.

Study materials for training are usually compiled and provided by an MSL trainer, an MSL preceptor (see below), or an MSL director. Every medical science liaison should receive a binder or electronic storage system containing the most relevant papers in the therapeutic area and/or about the product, and other background information.

All new medical science liaisons should be assigned a preceptor. A preceptor is an experienced MSL who can provide peer support in partnership with the MSL trainer and the MSL manager/director. A preceptor helps ensure that the newly hired MSL reviews and comprehends core training materials. During onboard training, the preceptor generally reviews informal processes, culture, chain of command/ hierarchy, and meeting policies and priorities. The preceptor should also assist the new MSL to become familiar with key resources, tools, and processes, including:

- Medical science liaison competencies
- Time cards, project-labor reporting
- Electronic document storage
- Review of key clinical data (if applicable)

- Introduction to key internal stakeholders and key meetings (with manager/admin support)

The preceptor should participate in an initial meeting with the new MSL, trainer, and manager/director, soon after the beginning of employment, to discuss:
- The training program
- Roles and responsibilities of the support team
- Expectations for the first ninety days

Ideally, the preceptor will maintain weekly contact with the new MSL for the first three weeks, then monthly, as needed, for one year.

Although many options exist for the design of MSL training programs, and a varied approach to training maintains interest and enthusiasm, it is important to keep in mind that continued training is vital to a competent and successful MSL.

Chapter 3:
Investigator-Sponsored Trials and the Medical Science Liaison

In the United States, pharmaceutical and biotechnology companies sponsor the majority of clinical research. However, with sufficient support and approval, any physician can decide to become a "sponsor" (otherwise known as a sponsor-investigator) and devise a trial protocol to study the efficacy and safety of an investigational therapy administered to humans. This kind of study is called an investigator-sponsored trial (IST).

ISTs are generally supported with funding and study drug supply provided through grants from research institutions and biotech/pharma companies. However, ISTs are distinctly separate from a company's clinical development (Phases I, II, and III) program for a molecule. This brings us to an important distinction: unlike industry-sponsored clinical trials—where the "sponsor" is the company that funds the trial—in the case of ISTs, the "sponsor" is not the company providing support, but rather the physician, who assumes legal responsibility for overseeing the study. Further, in an IST, it is this physician (i.e., the sponsor-investigator), not the funding company, who develops the protocol. In short, ISTs are initiated by one or more investigators who have a specific idea and take ownership of the study.

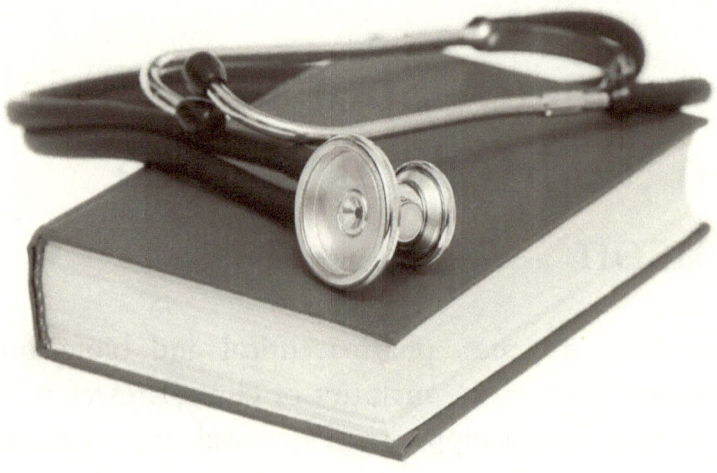

In their role as biotech/pharma representatives who interact routinely with opinion leaders in their day-to-day work, MSLs are uniquely positioned to facilitate the IST process and aid in the progression of such studies. The MSL works with the investigator to ensure the study has all the necessary elements necessary to be administered appropriately, in addition to making sure the study is moving ahead on time, and within budget.

All sponsors must comply with the FDA regulations. Sponsors are responsible for identifying qualified investigators, submitting and maintaining an Investigational New Drug application (IND) and related documentation, and ensuring that the FDA, all participating investigators, and regulatory agencies are promptly informed of significant adverse effects or safety risks with respect to the investigational therapy. The sponsors are, furthermore, obligated to provide all sites with the information they need to properly conduct the investigation, and to ensure that the study is properly monitored and conducted in accordance with the protocol and with all applicable regulations, rules, and guidelines.

The responsibilities of a sponsor may be contrasted with those of a (non-sponsor) investigator, the latter of which is defined as any physician who assumes full responsibility for the treatment and evaluation of patients enrolled in research protocols, as well as the integrity of the research data.

Many trials involve collaboration among many investigators, one of whom is designated as the principal investigator. In the case of an IST, the principal investigator and sponsor-investigator are usually the same person. The primary responsibilities of a principal investigator are set forth in 21 CFR, Part 312, Subpart D—"Responsibilities of Sponsors and Investigators":

- Obtain initial and ongoing Investigational Review Board (IRB) review in compliance with the FDA requirements in 21CFR 56. (An institutional review board [IRB], also known as an independent ethics committee [IEC] or ethical review board [ERB], is a committee that has been formally designated to approve, monitor, and review biomedical and behavioral research involving humans with the alleged aim to

protect the rights and welfare of the research subjects, (http://www.fda.gov/oc/ohrt/irbs/default.htm).
- Obtain the informed consent of subjects.
- Conduct and personally supervise the study according to protocol.
- Prepare and maintain accurate case histories.
- Maintain adequate records of drug accountability.
- Report clinical findings and adverse reactions in accordance with FDA requirements.
- Maintain all documentation and patient data information.
- Communicate with all personnel involved in study coordination.

The principal investigator may also designate sub-investigators who will be assisting in the conduct of the investigation by listing them on FDA Form 1572. If a sub-investigator is located at a different institution, the sub-investigator must submit the protocol and patient informed consent form to his/her respective IRB. Like the principal investigator, sub-investigators must have demonstrated competence in the treatment of appropriate patients as defined by the research sponsor and have the ability to enroll a minimum number of patients as established by the research sponsor.

IST sponsor-investigators must submit proposals in order to receive grants for funding and/or drug supply. Companies that fund ISTs typically have a policy outlining the documentation that must be included in IST proposals and the process for submitting these proposals to an in-house review committee. The review committee is usually composed of individuals from multiple functional areas of the company, including Medical Affairs, Regulatory Affairs, Clinical, Legal, and Supply Chain. Companies generally prefer to award grants for clinical research relevant to its therapeutic diseases of interest. In many instances, the goal of the IST program is to provide clinical data that complements corporate sponsored studies regarding the optimal use of their molecules, at the same time, providing clinical benefit to patients.

Companies that support investigator-sponsored trials must ensure that any IST grants, for funding, study drug, or other resources, are awarded in a consistent manner and are compliant with all laws and regulations, as well as legal, regulatory, and fraud and abuse guidelines. The *Code of Federal Regulations* addresses this. For example, 21 CFR, Part 312.7, refers specifically to requirements concerning the promotion and charging for investigational drugs. Specifically, charging investigators a fee for an investigational drug being studied under an IND is not permitted without the prior approval of the FDA. Consequently, all clinical trials, including ISTs, that involve investigational (non-commercially available) products should receive drug supplies for the study at no charge.

The committee reviewing the grant may consult the MSL for guidance on the acceptability of the proposed grant request. The size of a grant is determined by multiple factors, including study design and complexity, type of supportive care, accrual projection, research impact, and degree and duration of patient assessment, including the length of the patient follow-up period.

The appropriate size of a grant for an IST should be modest and reflective of study-associated costs, such as:
- Data management
- Secretarial support
- Statistical consulting expenses

Grants for ISTs are not usually paid out in one lump sum. In many instances, the total grant is divided into several payments based on the achievement of predetermined commitments, or milestones, during the conduct of the trial. Each IST may have different milestones stipulated. Examples of milestones include:
- IRB approval of the protocol and patient consent form
- 25 %, 50 % 75% and 100% patient accrual
- Publication of clinical trial results

By providing the necessary application forms and informing the sponsor-investigator of the company's proposal review criteria (which may also be printed on the submission form), the MSL can help ensure that the IST proposal is well conceived and organized, and that it complies with ICH/FDA/OIG Safe Harbor guidelines, thereby enhancing the probability that it will receive committee approval.

Although each company has its own processes, a typical proposal consists of the following items:
- Letter of intent
- Concept or draft protocol
- Proposed budget
- Curriculum vitae of principal investigator

The letter of intent (LOI) serves as the preliminary step for communicating a research concept. The LOI communicates the physician's willingness to develop and implement a clinical research protocol. The LOI should be brief (one to two pages) and include a number of elements, including the study objectives, background and rationale for the study, sample size, inclusion/exclusion criteria, study drug dose and method of administration, length of treatment period, criteria for evaluating data, monitoring plan, and budget. The MSL may communicate this list of elements to the investigator to help ensure that submitted LOIs are complete.

If the proposal seems feasible, and after initial review of the LOI by the company's review committee, and the next step is protocol development. The sponsor-investigator writes the protocol, but can refer to a template, if available. Protocol authors should carefully review the company's investigator brochure (IB) as they develop their protocol, to ensure they have listed all known toxicities and included all reasonable measures to monitor any toxicities in their study.

The MSL should ensure that the final IST protocol contains the elements and meets the criteria suggested by FDA and by the company. In general, every IST protocol should address or include the following:

- Title of proposed study and table of contents
- Background and rationale for the study
- Study objectives
- Study design (e.g., randomized, double-blind, etc.)
- Number of patients and groups
- Inclusion/exclusion criteria

- Treatment plan (e.g., dosage and duration)
- Dose modification (i.e., how and when the dose may be modified)
- Efficacy endpoints
- Safety endpoints, including a description of how patient safety will be monitored and adverse events reported
- Criteria for terminating treatment and/or removing a patient from the trial
- Statistical analysis plan
- Data management
- Early termination procedures (if appropriate)
- References
- Appendices should include copies of case report forms (the tool used to capture study data), as well as any tools used to measure outcomes in the study, toxicity scales, etc.

Some companies collect a final, IRB-approved version of the patient consent form from the investigator and house it in the IST files.

Part of the protocol approval process involves determining whether the IST needs to be conducted under an Investigational New Drug application. To assist the investigator in determining whether this is necessary, the MSL can consult the company's Regulatory Affairs representative. The MSL may also facilitate the investigator's IND submission by working with Regulatory Affairs to ensure that a letter is filed with the FDA authorizing the sponsor-investigator's IND to cross-reference the company's own IND for the investigational product. A copy of this letter is generally furnished to the investigator and becomes part of the IND application. The MSL may also refer the sponsor-investigator to the FDA Web site, or can assist the investigator in securing the appropriate forms needed to file an IND by visiting the FDA Web site (www.FDA.gov).

In some companies, the MSL obtains documentation from the sponsor-investigator confirming that the IND has been filed, such as a copy of the FDA letter confirming receipt of the IND.

MSLs may be involved with other procedures associated with IST management and facilitation, such as:

- Acknowledgment of IST proposal receipt
- Notification of committee approval of the IST
- Coordination with the company's legal department to establish appropriate grant agreements
- Distribution of Investigator Brochures (IB)
- Assistance with IST trial awareness programs
- Ensuring the required documentation is in place to authorize the initial shipment of drug supply to the IST site
- Reminding investigators of requirements with regards to serious adverse event notification
- Obtaining study closure documentation
- Following up on publication

All investigators evaluating non-FDA-approved drugs are required by good clinical practices to have the most updated Investigator Brochure from a manufacturer or sponsor. In many, if not most situations, the investigator's IRB will request a copy of the brochure. In addition, IST investigators who have received an IB from the funding company are sometimes issued updated drug safety information on a periodic basis, particularly IB updates and any new information that changes the safety profile of the drug.

Pending the execution of confidentiality agreements, as necessary, the MSL may facilitate the distribution of an IB to investigators who have submitted successful IST proposals. The MSL may also assist in furnishing IST investigators with drug safety updates, as available, and helping to address questions that investigators may have.

Definitions: Adverse Event and Serious Adverse Event

Although the process whereby an investigator reports adverse events is detailed in the study protocol, and the requirement for safety reporting may also be outlined in the investigator's contract with the company, the MSL may choose to review the safety reporting procedures with the investigator. The MSL should be available for all questions regarding drug safety reporting and should be familiar with all procedures for reporting.

An adverse event (AE) is any unfavorable and unintended sign, symptom, or disease, possibly associated with the use of an investigational (medicinal) product. An adverse event is considered a serious adverse event (SAE) if it:
• Results in death
• Is life-threatening (the term "life-threatening" in the definition of "serious" refers to an event in which the

patient was at immediate risk of death at the time of the event; it does not refer to an event that hypothetically might have caused death had the event been more severe)

- Requires hospitalization, or prolongs existing hospitalization
- Results in persistent or significant disability or incapacity
- Requires medical intervention to prevent one of the outcomes listed above
- Results in a congenital abnormality or birth defect
- Is medically significant

The definition of an SAE encompasses events that are both expected (e.g., in the FDA-approved product labeling or identified in nature, severity, or frequency in the current Investigator Brochure) and unexpected (e.g., not in the FDA-approved product labeling or Investigator Brochure). Any unexpected SAEs that are drug related must be also reported to the FDA in a timely manner, as described in 21 CFR 312.32. Investigators must also report unexpected and drug-related SAEs to their governing IRB or IEC as required by local regulations and guidelines.

Medical and scientific judgment should be exercised in deciding whether expedited reporting is appropriate in other situations, such as important medical events that may not be immediately life-threatening nor result in death or hospitalization, but which may jeopardize the patient. (Examples of such events are blood dyscrasias or convulsions.)

The investigator and his/her appropriate designees must follow federal regulations and IRB guidelines for reporting SAEs, regardless of whether they are expected or unexpected. Drug-related SAEs must also be reported to the investigator's Institutional Review Board or independent ethics committee. If the investigator is chairing a multicenter study, the event must also be reported to the participating subsites. If a subsite participating in a multicenter trial encounters an SAE, it must be reported to the principal investigator.

In addition to the investigator's safety reporting obligations to the FDA, the company may, likewise, require the investigator to notify its drug safety department of the event, using certain procedures or forms.

Facilitation of Study Progression

The MSL's responsibility is to identify the final outcome of all ISTs followed and to ensure proper documentation of this outcome in the company's IST file. When an investigator informs an MSL that his/her study has closed, the MSL must follow the relevant company procedures, based on the study outcome.

For example, if the IST was terminated prematurely or study accrual was closed before the target sample size was met, the MSL should obtain a copy of the letter from the investigator to his/her IRB explaining the reason for this change, any follow-up correspondence from the IRB, and a study summary explaining the reason for premature closure and any preliminary results. If the study was completed and the study met its accrual goal, the MSL should obtain a study summary for the company files, as well as a copy of any abstract or manuscript produced.

When an MSL learns that an IST investigator plans to publish or present study findings, he or she should follow the company's internal policies for review of IST publications and presentations.

Regardless of whether the study was completed or closed/terminated prematurely, study summaries should include the following elements:
1. Protocol title
2. Number of patients enrolled
3. Safety and response rates
4. Implications of the data
5. Reason for the premature study closure (if applicable)
6. Any publications derived from the IST should be cited

Upon completion of final study milestones, the MSL should also ascertain that the final grant payment was delivered.

Conclusion

The role and the functions of the MSL vary depending on the company's strategic plan and needs of the organization. Some MSLs specialize in certain kinds of activities. The MSL fills an essential role for biotech/pharma companies, because they handle requests for information from so many sources, including insurance companies, formulary decision makers, managed care, government payers, health care professionals, and others. The medical science liaison is the industry link between research communication and dissemination of scientific information.

> Pharmaceutical companies employ medical science liaisons (MSLs) to serve as information providers to clients and potential clients—from insurers to doctors in a medical group.
> The demand for MSLs has grown along with the legal and regulatory requirements pharmaceutical companies must satisfy.
> MSLs are not sales representatives. They have medical or scientific backgrounds, and so they can provide

peer-level interaction in a way that a business person can't. Typically, MSLs have a Ph.D. in the sciences (e.g., nutritional science), a pharmacy degree (PharmD) or an MD. Before becoming a liaison, they often get their start working in medical research or in hospital pharmacies.
- *quoted from CNN.com (http://money.cnn.com/galleries/2007/news/0702/gallery.jobs_in_demand/index.html)*

MSLs will continue to play and important part in the pharmaceutical/biotech industry. With the advancement in complicated medicines and therapies for various disease states, the educational needs of health care decision makers will increase. Likewise, the communication and educational needs of biotech/pharma customers will amplify, thus underlining the importance and value of the medical science liaison. In particular, MSLs are playing a growing role in small pharma and biotech companies, where the MSL not only provides education, but also often introduces the company to thought leaders and the market dynamics.

The medical science liaison role that was born almost forty years ago, continues to flourish and evolve. The MSL works on the company's marketed products as well as product pipelines to help develop therapies offering the potential to change the course of medicine for future growth. The MSL offers critical expertise and support to many areas of the pharma/biotech industry—research, education, clinical trial facilitation, to name a few. Most importantly, MSLs contribute to the noble cause of making a difference in patients' lives.

MSL-Specific Web Resources:

Pharm, LLC—www.pharmllc.com

MDea—www.mdeany.com/home.html

The Medical Affairs Company—www.themedicalaffairscompany.com/index.html

MSL Institute—www.mslinstitute.com

Science Oriented Solutions—www.medicalaffairs.com

Scientific Advantage—www.scientificadvantage.com, http://www.medicalscienceliaisons.com

About the Author
Susan E. Malecha, PharmD, MBA

Susan Malecha has over sixteen years of pharmaceutical/ biotechnology industry experience in medical affairs, new product development, managed markets, medical education, medical science liaison and management experience. She graduated with a BS in Pharmacy from Butler University, completed her Doctor of Pharmacy at University of Illinois at Chicago, and earned her Masters of Business Administration from Keller Graduate School of Management. She is currently is a Director for Field-based Medical Science Liaison teams for a large biotech company. She has also led field based medical groups at smaller biopharmaceutical company. Prior to her biotech positions, Susan was the Director of Managed Care field based medical team and the Director of Medical Education at a large pharmaceutical company, in addition to holding various positions at other mid-size and small companies. She has held adjunct faculty positions at University of Illinois at Chicago and Midwestern College of Pharmacy. She is an active presenter/lecturer on medical affairs topics for pharmaceutical industry. She participates on the Board of Visitors for Butler University College of Pharmacy and Health Sciences and on the Editorial Board for the MSL Institute. She has published papers in Pharmacotherapy, Drug Information Journal, MSL Quarterly, DIA Forum, and American Journal of Pharmaceutical Education. She has contributed to the published books, Erin Albert's "Medical Science Liaisons A to Z" and Broberg et al "Prescriptions to a Younger Self: What I learned after Pharmacy School."

Susan lives in San Diego with her husband and two children.

www.ingramcontent.com/pod-product-compliance
Lightning Source LLC
Chambersburg PA
CBHW021925170526
45157CB00005B/2191